YOUR KNOWLEDGE HAS VALUE

- We will publish your bachelor's and
  master's thesis, essays and papers

- Your own eBook and book -
  sold worldwide in all relevant shops

- Earn money with each sale

Upload your text at www.GRIN.com
and publish for free

The GlobalMode5 Vortex Coil. Fluid Flow in Spiral Vortex Structures

Michel Felgenhauer

Bibliographic information published by the German National Library:

The German National Library lists this publication in the National Bibliography; detailed bibliographic data are available on the Internet at http://dnb.dnb.de.

ISBN: 9783346874771
This book is also available as an ebook.

© GRIN Publishing GmbH
Trappentreustraße 1
80339 München

Print and binding: Books on Demand GmbH, Norderstedt, Germany
Printed on acid-free paper from responsible sources.

The present work has been carefully prepared. Nevertheless, authors and publishers do not incur liability for the correctness of information, notes, links and advice as well as any printing errors.

GRIN web shop: https://www.grin.com/document/1356563

The globalMode5 Vortex Coil
Fluid flow in spirally vortex structures

Michel Felgenhauer, Berlin Germany in Mai 2023

Fluid-mechanical vortex coils arise as spirally arranged coherent vortex filaments. One can assign physical properties to fluid-mechanical vortex coils, but formally there is no generally valid theory about spiral vortex formations in fluid mechanics. Theoretical key statements about vortex filaments have been known for a long time; the most important are from Helmholtz. A common feature of some modern theoretical approaches to ordered vortex configurations is that they designate coherent vortex formations as connected domains of dominant vortex strength.

For the phenomenology of "multiple (n) fluid-mechanical vortex coils" (global mode n vortex coil) presented here, Helmholtz's vortex filament theory, which is considered to be reliable, is first applied to Lagrangian coherent structures and expanded by an approach to the inner milieu of the vortex filaments.

Structures of this kind form systems that are capable of momentum induction, which in turn organizes the field at rest. For the presence of well-grouped vortex filaments, there is a conjecture about the self-organization (autopoiesis) of vortex filaments in a flow field.

Keywords: field, fluid mechanical vortex coils, vortex filament theory, Lagrangian coherent structures, vortex filaments, autopoiesis

<u>From the content</u>

Spiral shape

The multiple, fluid-mechanical vortex coil remains a synthetic construct. Vortex coils are fluidic compositions of coherent vortex filaments with ascribed properties. The physical properties of the vortex filaments derive from their well-defined internal milieu. Spiral vortex formations occur in nature, but they are "made"! They are synthetic in the sense that generating systems exist that generate fluid-mechanical vortex coils.

So far, Lagrangian coherent vortex systems have not had a universal definition in fluid mechanics. This also applies to spiral systems. The first formulations of spiral rotationally coherent vortex systems came from observations of particular fingering of the wing tips of land-soaring birds. The descriptions of multiple vortex coil systems had no relation to the natural habitats of the birds, but were only arranged laboratory experiments in the wind tunnel; it was as if the decoding of bird flight had begun in a hall in the mid-1980s. At least that's how it was seen at the Department of Bionics and Evolutionary Technology at the Technical University of Berlin. Later, technical airfoil models were examined in the wind tunnel and their suitability for generating spiral vortex systems.

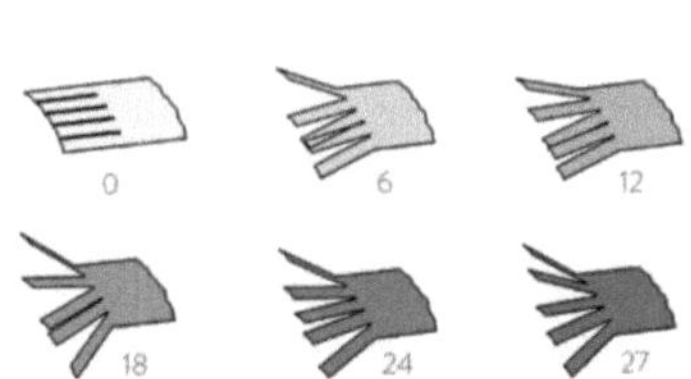

Fig.1: Conditioning of the segmented tip contour of a model wing. Lift L and drag W were measured. After only 27 variation and selection campaigns, the synthetic wing resembles its biological model: the plumage of the land-soaring bird[1].

At the beginning of the investigations by Nachtigall (Saarbrücken) and Rechenberg (Berlin), little was known about the very special fluid mechanics of the vortex coil structures and the theory required for this. In Berlin, the first technical laboratory model of the fingered bird's wing was a splitted edge curve contour of a formerly compact model wing that was now slotted at the end of the wing. At that time, all experimenters in Berlin worked primarily with sheet metal (Fig.1), because lead surfaces offered certain creative freedom when constructing fingered model wings.

With an optimization strategy tailored to wind tunnel tests, the glide ratio $(cL/cW)^2$ of a slotted and fingered lead wing was improved by about 10% at the

---

[1] Based on illustrations from the pool of the department of bionics and evolutionary technology at the Technical University of Berlin; see also: Stache, M.: Evolutionsstrategisches Design von Tragflügelspitzen. DGLR- Jahrbuch, DGLR-2006-200, 2006, pp. 1131-1138

[2] Gleitzahl eines Tragflügels, Kennwert aus der Aeromechanik (Lilienthal-Beiwert) und berechnet aus gemessenem oder berechneten Widerstandskoeffizienten cW und dem Koeffizienten des fluiddynamischen

TU Berlin in the 1990s compared to its compact initial geometry. Only the preparation of the stork's wing behaved cheaper in the wind tunnel; with its exorbitantly low drag coefficient.

In the case of the lead wing model (Fig. 1), the partial wings or wing tips of the feather fingers form a curved chain of source points. For a long time, this arc-shaped arrangement of the source points was considered less productive for wake flow and momentum exchange there. In fact, the geometry of the vortex-generating system (ellipse or arc) has a great impact on the quality of the vortex-spiral system.

In the 1980s, winglets[3] entered the stage of research institutes and a short time later of the development departments of the aviation industry. Winglets are fixed partial wings on aircraft wings.

Let's briefly discuss which research questions and research results in the 1990s and 2000s advanced the development of aerodynamically approved wings that could actually be used commercially in civil aviation.

The work of Tucker (1993)[4] is fundamental to the study of the biological system. The ellipse configuration of biological fingers is not recognized here. The model investigations are carried out with geometries according to the arch configuration. Smith and Komerath[5] et al. publish 2001 results from wind tunnel tests in which the vortex cores of the multiple winglets form a chain (arc configuration). Entz and Correa[6] et al. find an improvement in lift/drag performance of 12% to 14% for the three-finger case over the closed tip arch on a model wing (arch configuration). A numerical analysis of the triple configuration is described by Thimmegowda[7] et al. Likewise, arc configurations are investigated by Sevillano[8] et al. for very small flight aggregates. The vortex system behind

---

Auftriebs CL (Lift). Im Falle des antriebslosen Gleitflugs eines Vogels oder eines (Flug-) Modells entspricht die Gleitzahl zugleich dem Verhältnis aus zurückgelegter Wegstrecke und Höhenverlust.

[3] Winglets (wörtlich: englisch Flügelchen) bzw. Sharklets (Bezeichnung für Winglets bei Airbus), deutsch Flügelohren[1], sind meistens nach oben und seltener nach oben und unten verlängerte Außenflügel an den Enden der Tragflächen von Luftfahrzeugen. Sie sorgen für eine bessere Seitenstabilität, verringern den induzierten Luftwiderstand und verbessern so den Gleitwinkel sowie die Steigzahl bei niedriger Geschwindigkeit. https://de.wikipedia.org/wiki/Winglet

[4] VANCE A. TUCKER (1993) GLIDING BIRDS: REDUCTION OF INDUCED DRAG BY WING TIP SLOTS BETWEEN THE PRIMARY FEATHERS. J. exp. Biol. 180, 285-310 (1993)

[5] M. J. Smith·, N. Komerath+, R. Ames·, O. Wong·, (2001) PERFORMANCE ANALYSIS OF A WING WITH MULTIPLE WINGLETS; School of Aerospace Engineering, Georgia Institute of Technology, Atlanta, Georgia and J. Pearson, Star Technology and Research, Inc., Mount Pleasant, South Carolina.

[6] Cosin, R. , Catalano, F.M. , Correa, L.G.N. , Entz, R.M.U. (2010) AERODYNAMIC ANALYSIS OF MULTI-WINGLETS FOR LOW SPEED AIRCRAFT Engineering School of São Carlos - University of São Paulo.

[7] Hariprasad Thimmegowda (2016) Computational and Experimental Analysis of Multi-Winglet at Low Subsonic Speed Conference Paper · August 2016

[8] A. A. Rodríguez Sevillano *, R. Bardera Mora **, M.A. Barcala Montejano*, E. Barroso Barderas ** and I. Díez Arancibia *. (2019) Design of Multiple Winglets for Enhancing Aerodynamics in a Micro Air Vehicle. 8TH EUROPEAN CONFERENCE FOR AERONAUTICS AND SPACE SCIENCES (EUCASS)

winglet configurations is studied by Zang, Wanng and Fu[9]. Executed constructions are examined by Ning and Kroo[10] and Merryisha and Rajendran[11] , as well as by Scholz[12] for comparison. An experimental comparison between a single and a multi-winglet is made by Balagurumurugan et al[13]. A numerical model (Fluent) of a mono-winglet is described by Abdelghany et al[14]. Bird wing-inspired winglets (arc configuration) investigate Hossain, Rahman et al[15]. Also executed constructions examined Putro and Pitoyo, et al[16]. The last publication of an arc configuration from the Berlin department of bionics and evolutionary technology comes from Stache (2006)[17]. It therefore seems permissible to clarify the state of international science in the late 2010s with the arc configuration of the generator system. Recently, a study on biology-inspired multi-winglets by Yussof et al[18]. (2022) with a 7-fold fingering on the edge arch of a model wing, also in arch configuration.

To date, we only know the impulse effectiveness of spiral arrangements from the few vortex structures that have been experimentally investigated and measured in wind tunnels with precisely this question in mind. Also, we do not know whether the fluid mechanical effects are scalable or not. It can be observed that the impulse induction of Lagrangian coherent vortices in the flow field is cumulative and thus appears to be a conservative phenomenon. Compensations often occur in cumulative processes: physical effects cancel each other out. In this

[9] Zang, Wang und Fu (2019) GENERATION MECHANISM AND REDUCTION METHOD OF INDUCED DRAG PRODUCED BY INTERACTING WINGTIP VORTEX SYSTEM. School of Aerospace Engineering Tsinghua University, Beijing, China

[10] Andrew Ning, Ilan Kroo (2010) Multidisciplinary Considerations in the Design of Wings and Wing Tip Devices. Brigham Young University – Provo, and Stanford University

[11] Samuel Merryisha1, Parvathy Rajendran (2019) Review of Winglets on Tip Vortex, Drag and Airfoil Geometry School of Aerospace Engineering, Universiti Sains Malaysia, Engineering Campus. Journal of Advanced Research in Fluid Mechanics and Thermal Sciences 63, Issue 2 (2019) 218-237

[12] Scholz, D. (2018) Definition and discussion of the intrinsic efficiency of winglets, Aircraft Design and Systems Group (AERO), Hamburg University of Applied Sciences; Aerospace Europe CEAS 2017 Conference, 16th-20th October 2017, Palace of the Parliament, Bucharest, Romania, Technical session Aircraft and Spacecraft Design

[13] R. Balagurumurugan, A.Yadav, A. Ahmed R, S. Narayanan S (2016) Experimental study of single and multi-winglets. Article in Advances and Applications in Fluid Mechanics · April 2016.

[14] E. S. Abdelghany , E. E. Khalil, O. E. Abdellatif and G. ElHarriri (2016) WINGLET CANT AND SWEEP ANGLES EFFECT ON AIRCRAFT WING PERFORMANCE in Proceedings of the 17 MP 258 th Int. AMME Conference, 19-21 April, 2016

[15] Altab Hossain, Ataur Rahman, A.K.M. P. Iqbal, M. Ariffin, and M. Mazian (2011) Drag Analysis of an Aircraft Wing Model with and without Bird Feather like Winglet, World Academy of Science, Engineering and Technology, International Journal of Aerospace and Mechanical Engineering. Vol:5, No:9, 2011

[16] S H S Putro, B J Pitoyo, N Pambudiyatno, Sutardi, and W A Widodo (2020) Comparison of the winglet aerodynamic performance in unmanned aerial vehicle at low Reynolds number. IOP Conf. Series: Materials Science and Engineering 1173 (2021) 012002 2

[17] Stache, M.(2006): Evolutionsstrategisches Design von Tragflügelspitzen. DGLR- Jahrbuch, DGLR-2006-200, 2006, pp. 1131-1138.

[18] Hamid Yusoff, Koay Mei Hyie,*, Halim Ghaffar, Aliff Farhan Mohd Yamin, Muhammad Ridzwan Ramli, Wan Mazlina Wan Mohamed, Siti Nur Amalina Mohd Halidi (2022): The Evolution of Induced Drag of Multi-Winglets for Aerodynamic Performance of NACA23015, Journal of Advanced Research in Fluid Mechanics and Thermal Sciences 93, Issue 2 (2022) 100-110.

way, an inducing system can couple momentum into the field at the body-fixed, Lagrangen level without this production becoming visible at the Euler level.

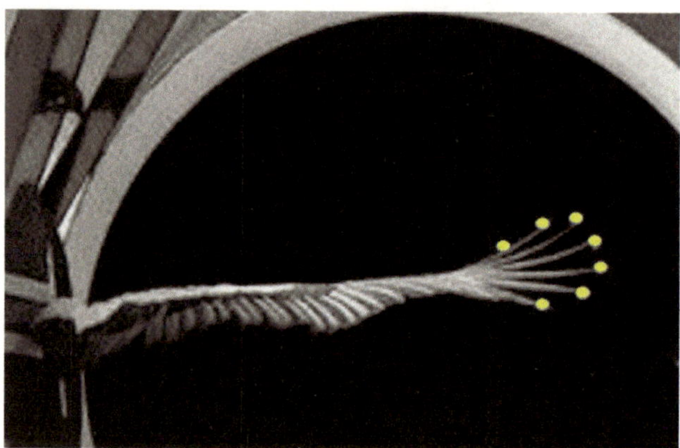

Fig.2: Specimen of the stork wing in the wind tunnel under flow load. The seven source points at the feather finger tips are marked in yellow.
Photography: Fachgebiet Bionik und Evolutionstechnik, Technische Universität Berlin (um 1995).

Vortex generators were able to continuously synthesize more or less compact, but in any case coherent vortex filaments and release them into the flow. Passive ordering processes were involved in what happened. It was observed that neighboring vortex threads influence each other and, immediately after their formation, begin to rotate around a common center, even to literally "dance with each other". This dance is optimal for an elliptical configuration of vortex filament sources and forms a compact, shell-shaped vortex filament coil downstream.

Downstream from the wind tunnel and in the wake of the perturbation contour, the spiral vortex structure remains stable. In the interior and along this helically wound vortex tube, the flow actually remains just as "rotor-free" as predicted by the widely unpopular potential theory. $v\infty$ is the fluid velocity at the mouth of the open wind tunnel and thus the so-called "apparent fluid velocity" at the specimen to be examined. Inside the vortex coil, the flow velocities take on a multiple of the amplifying flow $v\infty$.

At that time efforts were being made to find a theoretical model. In his dissertation (1988), Peintinger proposed a method for calculating fluid-mechanical vortex coils according to Biot and Savart's law[19], a formula derived

---

[19] Benannt wurde dieses Gesetz nach den beiden französischen Mathematikern Jean-Baptiste Biot und Félix Savart, die es 1820 formuliert hatten.Es stellt neben dem ampèreschen Gesetz eines der Grundgesetze der Magnetostatik, eines Teilgebiets der Elektrodynamik, dar. https://de.wikipedia.org/wiki/Biot-Savart-Gesetz
The Biot–Savart law is fundamental to magnetostatics, playing a role similar to that of Coulomb's law in electrostatics. When magnetostatics does not apply, the Biot–Savart law should be replaced by Jefimenko's equations. The law is valid in the magnetostatic approximation, and consistent with both Ampère's circuital law

from general field theory. Unfortunately, the method does not go beyond the doctrine of that time and the one hundred and seventy years that followed, which originated from the axis-conform calculation, and was only mentioned once as a theoretical model of fluid-mechanical vortex structures (Kaschub 1988).

The biological wing is a complex structure and carrier of different physical effects. So what is the typical thing about a bird's wing? Every flying system has a set of basic design ideas: high performance, low weight and the aerodynamic shape. For a safe flight, the senses, especially vision, must be sharp. Birds possess excellent eyesight, perhaps the most powerful of any vertebrate. The entire physique of a bird is adapted to the flying way of life. His bone system is a lesson in lightweight construction. Another weight loss adjustment is the absence of some organs. In the course of evolutionary optimization, today's modern birds are toothless and have no muscular jaws. The bird's beak is an adjustment that reduces the weight significantly.

Soaring bird in the wind tunnel

The first investigations into fluid-mechanical vortex coils took place in the laboratory. The wind tunnels and measuring devices of the TU Berlin were suitable for carrying out quantitative investigations into the air resistance and the dynamic lift of exhibits of different bird wings. The traditional tasks and methods in the wind tunnel included determining the vorticity of the flow being examined. In the 1980s, this was only possible qualitatively at the measuring facilities of the TU Berlin. However, spatial maps of the emergence of gradual vorticity in the field have been made. This work result from vortex probe analyzes crystallizes for the first time the spatial form of spiral systems in the wake of a bird's wing flow: The observed vortex areas form (i) stationary structures; they are (ii) discrete, distinct from the surrounding flow, and the domains appear (iii) as threads to be connected (coherence). The vortex filaments are (iv) ordered into a radially symmetrical spiral, which apparently happens by itself. Later it was found that these vortex threads (v) have an astonishing stability, and even more: bundles of these vortex threads are stable against external disturbances. However, what is phenomenal about these vortex coils is (vi) that the container flowing off behind the interfering contour rotates! In the center of the vortex spiral, (vii) a higher fluid velocity is registered than in the inflow boundary condition $(v_{COIL}/v_\infty)>1$ of the wing.

Wind tunnels have long been established and widespread as a simulation method and laboratory instrument. A wind tunnel is more of a building than an apparatus. Here the medium, in our case the air, moves towards the measurement object,

---

and Gauss's law for magnetism. It is named after Jean-Baptiste Biot and Félix Savart, who discovered this relationship in 1820.

flows around it and flows off. We measure selected physical variables on a stationary measurement object. If we examine the flight of birds, we cannot avoid recognizing that with a wind tunnel the principle of measurement and effect subject, i.e. the cause of a physical event and its effect on a biological flight system, is exactly reversed! The wind tunnel reverses the flow events to be described.

Let us now look at power- and work processes in a fluid at rest! A bird or a model airplane soaring through a fluid at rest works off a height potential in a very elegant way. In the still fluid (and also in the moving fluid) this process is a "work process" and means: The flight system couples energy into the field.
A wind turbine is completely different. The flow energy that a moving fluid transports depends on the density of the medium; in the case of air, the energy density is rather low. That is why windmills are usually large. The energy conversion is a "power process" here. The wind turbine's repeller decouples energy from the field. Windmills are power machines.
Obviously again, the propeller. As a power machine, a propeller couples energy into the field with the aim of generating reaction forces. The reactive forces become measurable as "thrust". From this perspective, the difference between a wind turbine and a hair dryer, is easy to understand. In fact, measurements in the plane of a wind turbine repeller show that the flow is decelerated here and flow momentum is transmitted: out of the field and into the machine!
Traditional theoretical investigation methods and modern numerical models of fluid mechanics (Computational Fluid Dynamics, CFD) examine exactly this. Immediately and effortlessly we find ourselves in the classic situation of a wind tunnel setup, in which a fluid flows around a body at rest. As a rule, control volumes are declared there, in which the "internal components" to be examined are located. A parameter of the initial boundary conditions is usually the velocity vector for $v\infty$ at the edges of the flow space; the pressure condition or a defined volume flow then prevails at the other system boundaries.

So let's follow the process of flying in the three-dimensional field. When driving a wing through a standing fluid, this fluid continuum experiences complicated deformations. The deformations come from the interaction of the fluidic space with the moving wing.
Driving means: clamped down on one side and finite; we imagine a wing profile, an upper and lower side and a tip area at the wing tip. From the perspective of the fluid and the initial boundary condition of a pristine field, the wing is an interference contour! Below this interfering contour, the fluid responds to the momentum input required for the deformation of the continuum with a downward movement. This downward movement of the fluid is local. It takes

place in the immediate vicinity of the interference contour. The moving wing carries this environment with it: it practically plows through the continuum! The term "downwash" has become established on the wing and for the observation of the phenomenon "downward movement of the fluid after impulse input". The term downwash[20] comes from the time when people began to study the flow field around helicopter propellers.

If we follow the argument about a moving interfering contour in a stagnant fluid, then kinetic energy is coupled into the field during the downwash as well as with the edge vortices from the flow around the finite contour. The kinetic energy comes from a movement cause and is lost for the driving system. The term "induced drag" has (unfortunately) been established around the energy loss from the tip vortices, in semantic differentiation from frictional and form drag forces of an interference contour moving in the fluid. The edge vortex and the fluid mass accelerated downwards from the wing consumes energy. The "induced drag of a tip vortex" is very probably an unfortunate term in the argumentation about tip vortices at a disruptive contour. In fact, "induction processes in a field" are always discussed when there is talk of the impulse effectiveness of a vortex structure. Unfortunately, however, there is no force induced in the field by a boundary vortex, as we know it along a line of force, for example. Nevertheless, Prandtl's calculation approach for the "induced drag" delivers such accurate results that even after a hundred years it is part of the repertoire of flow simulation.

In addition to the downwash, the "sidewash" is now known, a narrative that actually comes closer to the physical processes on the wing!

Corridor model

Without a perturbation, no flow will flow out of a control volume in a fluid at rest. In a mind game, we stake out a three-dimensional control area in a fluidic space, the field, so that a kind of "canal with standing fluid" is created: the corridor! This is where the following experiments will take place. As the first interfering contour, a wing of a model aircraft sails through the field in such a way that its wing tip glides through the control corridor. In this scenario, the flight system should not suffer any altitude loss. In flight: A continuous vortex track is now forming along the central axis of the corridor. The thread-like vortex track originates from the flow around the edge edges of the moving wing in the stationary fluid.

Let's underlay the mind game with the calculation results of a computer simulation based on field theory. In the numerical corridor model, a wing sails through a stationary field just such that its wing tip glides through the control room from end to end. A vortex track becomes visible (Fig.4. left in the picture).

---

[20] in der Luftfahrt bezeichnet Abwind (englisch downwash) einen technischen Abwind, wie er von Flugzeugen und Hubschraubern von den Tragflächen bzw. Rotoren bei der Erzeugung von dynamischem Auftrieb entsteht.

This one-dimensional vortex thread is coherent, energetic and dimensionally stable!

We know from vortices that their inner milieu is characterized by the vector vorticity ω [s-1] (vorticity ,)[21]. In the model presentation, the vortex thread is a one-dimensional (vortex) filament in the sense of Helm-Hotz' vorticity laws and a Lagrange coherent system (LCS) as described by Haller (Haller2000).

In a graphic for the numerical simulation of the corridor model and for the movement of an interference contour in a fluid at rest, vectorial flow quantities are perpendicular to the direction of movement (YZ plane, Fig. 3., middle) and along the vortex track (XY plane, Fig. 3., on the right in the picture). The specific induced impulse corresponds to the velocity induced in the field[22].

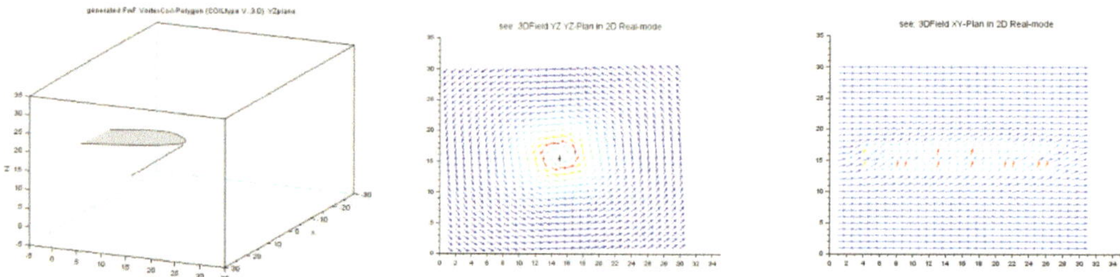

Fig.3: Numerical simulation of a vortex track in a fluid at rest. Schematic: wing and corridor model (left); Distribution of the velocity vectors in the YZ plane (middle) and in the XY plane, on the right in the picture. (nach: Felgenhauer (2023). Fluid-Filament- Wechselwirkung, FFW).

The vortex thread "organizes" the flow field! The diagrams in Fig.4 show a Y-Z section and the components of the Lagrangian velocity gradient in the axial direction of the vortex filament: u; radial components, perpendicular to the main direction of movement: v, w; and the resulting velocity r in the field. Fig.6 shows the calculation results in the YX plane of the vortex thread.

One result of the simulation is the visualization of the vectorial induced velocity vi in a sectional plane of the flow field. We see that the stretched vortex filament applies little velocity in the axial direction u and the radial component w of the

---

[21] In der Simulationspraxis wird die statt der Wirbelstärke ω [s-1] häufig die Zirkulation Γ [m2s-1] verwendet. Also: In continuum mechanics, vorticity is a pseudovector field that describes the local spinning motion of a continuum near some point (the tendency of something to rotate), as would be seen by an observer located at that point and traveling along with the flow. It is an important quantity in the dynamical theory of fluids and provides a convenient framework for understanding a variety of complex flow phenomena, such as the formation and motion of vortex rings. https://en.wikipedia.org/wiki/Vorticity

[22] Die Komponenten (u,v,w) der in das Feld induzierten Geschwindigkeit vi, die gleich dem vektoriellen spezifischen induzierten Impuls ji[ms-1] ist.

velocity dominates what happens along the core of the vortex filament in the field: the resultant r is fed almost entirely from the radial component. This means that the drag, which comes from the flow around the edge edges of the lift-generating wing, is primarily due to the radial flow, which is fanned by the driving system. This is exactly what the "SideWash" mentioned above as a metaphor is.

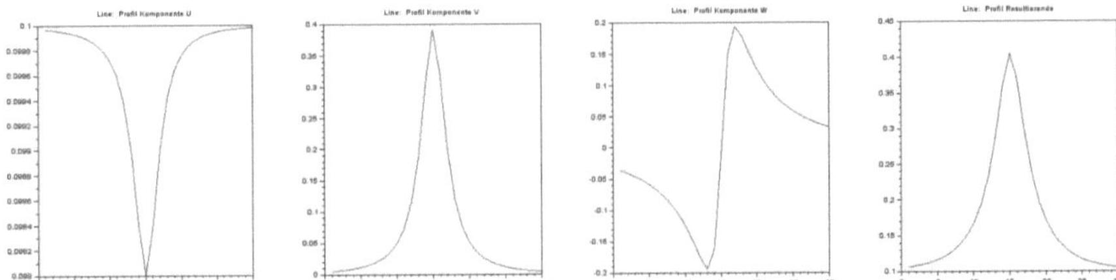

Fig.4: Y-Z section (X=15LE) and the components of the velocity gradients in the direction of the axis of the vortex thread: u; radial components, v, w; Resulting local velocity in the field: r. (nach: Felgenhauer (2023). Fluid-Filament-Wechselwirkung, FFW).

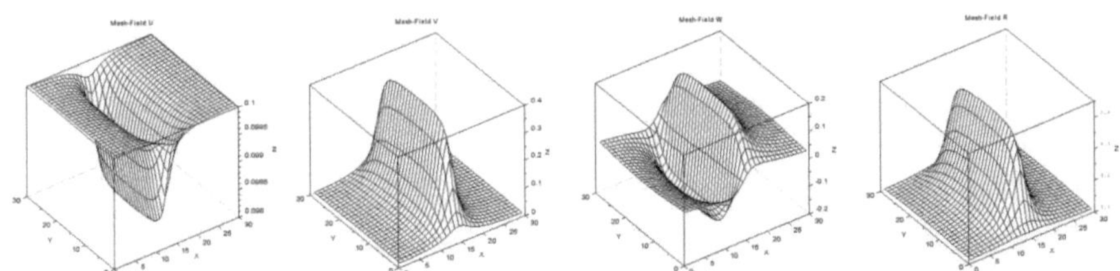

Fig.5: Section through the calculated field in the X-Y plane (Z=15LE). Impulse effect of an "elongated" vortex filament in a fluid at rest (Fig. 3); the components of the velocity gradient u, v, w; and the resultant r. (left to right; nach: Felgenhauer (2023). Fluid-Filament- Wechselwirkung, FFW).

Communicating vortex filaments

For a long time we did not have a working model of a fluid-structure interaction for a group of spatially distributed vortices. The complicated combinatorial fluid-filament interaction (FFWW) of the simulation model leads to a vortex coil system that organizes the field so that flows start to follow a physically based fluctuation. This appears unreal to the observer at first! How can it be that flow lines form "by themselves" in a field?

The numerical model assumes that a wing tip, splitted into five feather fingers - also encloses five identical circulations. Elsewhere it will be shown what non-homogeneous circulation distributions at a feathered marginal arc cause spiral formations. The oddness of the fanning plays no significant role in the simulation. The five-fold interaction scenario is discretized 5 times p-dimensional on the part of the vortex filament: Each finite section on the vortex filament contributes to the induction of a velocity and a specific momentum at each location in the field. The vortex sources are now finite small but localizable sections on each vortex filament. The very small sector of an inductively active vortex filament in Lagrange coordinates affects all locations in the field in Euler coordinates. The entire scenario of vortex threads has an inductive effect on every point in the field. This combinatorial balance is complex in the numerical simulation. The effects in the points of the field are cumulative.

The fluidic vortex coil (WSP) now considered is a spiral structure made up of five interaction partners: global-Mode5-WSP! The following applies to a 5-way interaction: If a moving or stationary fluid encounters flexible forms of moving interfering contours, these systems influence each other. In addition, they always shape the fluid surrounding them when the interfering contours are flow-inductively vital. If filaments manipulate other filaments inductively, one can speak of a fluid-filament interaction (FFWW).

The autopoietic, self-referential interaction of five eddy threads in a fluidic field means an a priori simultaneity of all fluidic partners that are impulse-effective in the field[23]. This includes interactions with neighboring structures that are able to shape each other. Each location on each vortex filament is a source point; every location on each vortex filament is a reference point. Near, but also far from any place on the vortex filament, each vortex source has an induction-effective influence "on itself".

The simulation tells us how the flow space, which is initially at rest, is set in motion and thus organized by the induction effects of the spiral vortex structure. As expected, the rotating vortex coil system system globaMode5WSP "pumps" a standing fluid through its interior: fanning from the left to the right (Fig.7); namely "rotor-free"!

The corridor contains the fluid at rest. A 5-fold vortex coil is created in this 3D corridor model successively and self-referentially (Fig. 6). We see that the vortex

---

[23] engl. autopoiesis, (von griech. autos und poiesis: Selbstherstellung) bedeutet die Hervorbringung von etwas aus sich selbst, etwa die Produktion eines Systems aus der Struktur der Elemente, aus denen es besteht. Der Begriff der Autopoiesis stammt ursprünglich aus der Systemtheorie (Maturana, Varela, von Foerster) Maturana, Humberto R. u. Francisco J. Varela (1980): Autopoiesis and cognition: The realization of the living. Luhmann, Niklas (1985): Die Autopoiesis des Bewußtseins. Soziale Welt 36. Foerster, Heinz von (1993): Wissen und Gewissen – Versuch einer Brücke. Frankfurt a. M. (Suhrkamp).

coil in the inductive interaction process "degenerates a little" with progressive iterations.

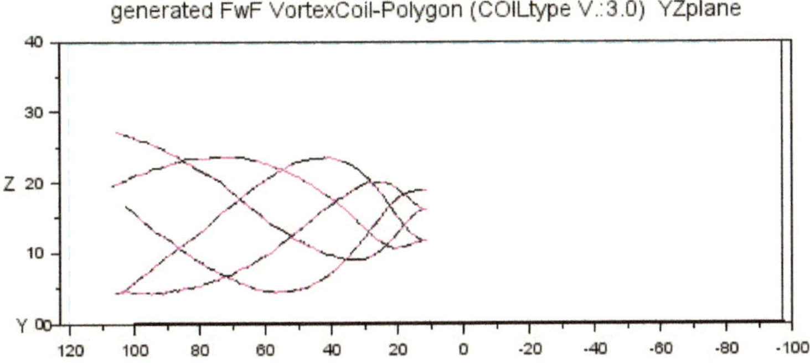

Abb.6: Corridor model of a 5-fold vortex coil: globaMode5WSP.

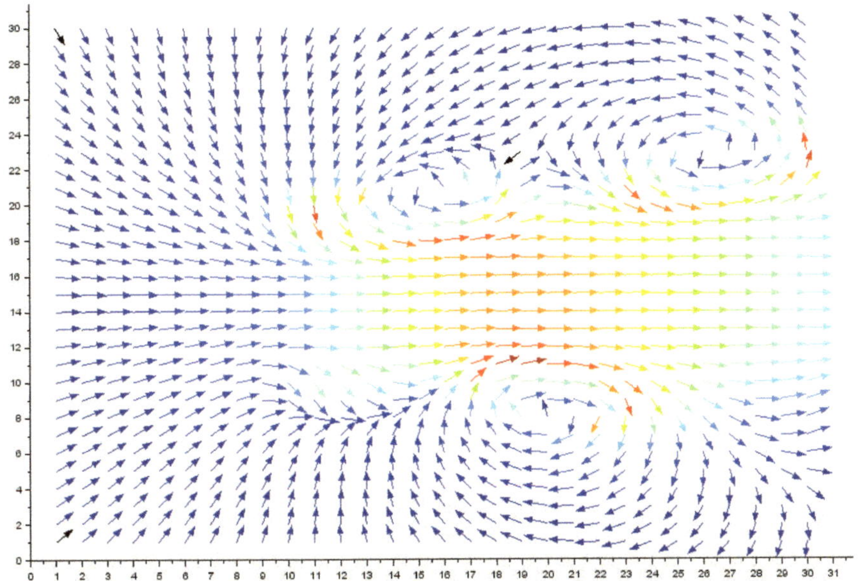

Fig.7: Flow pattern in the XY section plane (Z=15 LE)

Let us consider the fluid mechanical process again. In the setup of the model, the calculated vortex coil structure exists from a height of X=10LE, sucks in fluid from its surroundings, conveys matter along its central axis and out astern. If the vortex filaments are close together, for whatever reason, there is an intensive induction effect in the flow. This is an important finding for future design issues. To a bird, the fluid mechanics vortex coil is nothing less than a "biological jet propulsion!" To an engineer, it is a very elegant fluidic aggregate that generates reaction forces.

Conclusion

In the classic sense, the flow-mechanical vortex coil does not reduce the so-called induced drag of a lift-generating wing. However, it does add a thrust to the movement system that opposes the resistance. Induced drag is often referred to as the price to pay when ordering buoyancy: no free lunch! In this rhetoric, the impulse-inducing interaction events in and around a fluid-mechanical vortex coil would be the discount marks in the balance sheet. You get something back!

Summary. The gliding flight of a lift-generating unit through a fluid at rest leaves a vortex pattern in the fluidic space. Such a flow process, in which several vortex thread systems communicating with each other interact fluid-mechanically, so that momentum exchange prevails simultaneously at every point in a previously stationary space, has not yet been experimentally investigated for a stationary fluid.

The numerical experiment deliberately reverses the conditions prevailing in a wind tunnel: a moving interfering contour passes through a fluid at rest. Model calculations show that left to themselves vortex filaments in the wake of an interference contour begin to "dance with each other" in order to form a rotating spiral structure. This theoretical knowledge confirms assumptions from real observations and from measurement results on fluid-mechanical vortex coils in the wind tunnel.

Well, the functional structure of fluid-mechanical vortex coils is quite complex. And it's not explored in any way. With the theoretical models discussed here, it can be shown that in simulations of fluid-mechanical vortex coils, the current threads are compressed, analogous to a compression of magnetic field lines in an electric field. This was the prediction of the transfer of a general field theory to the fluid-mechanical concerns of a three-dimensional fluidic field.

The formation of a structure of fluidic vortex filaments left to its own devices is self-referential, autopoietic and interactive. An interaction model of interacting vortices is motivated by observations, fluid-mechanical investigations and measurements in wind tunnels (resting interfering contour, moving medium) and was successfully applied in this study to a stationary fluid with a moving interfering contour (corridor model).

The essay develops the principle and the formal requirements of the modeling and numerical simulation of five-thread fluid-mechanically effective vortex coil systems made of filaments against the background of self-referential fluid-filament interactions. A Ramsey permutation method is integrated into the numerical model as the algorithmic core of the calculation method for a flow field containing vortex coil systems. The result of the work is the simulation and calculation of the impulse induction through an interference contour in a fluidic space (corridor) at rest. The numerical model is to be further developed.

Michel Felgenhauer Berlin, Spring 2023

Michel Felgenhauer is the pseudonym of Michael Dienst from Wiesbaden, Germany. I live and work in Berlin and have been a lecturer in Bionic Engineering at the UDK Berlin and at the Industrial Design Institute Magdeburg since 1995.

Martha Felgenhauer died in 1943 as a young woman in Ziegenhals, Silesia. Those who knew her say, we are kindred spirits. So occasionally I tell my grandmother stories of the funny science..

Bibliography, sources and further reading

[Abbo-59]   Ira H. Abbott, Albert E. von Doenhoff: Theory of Wing Sections: Including a Summary of Airfoil Data. Dover Publications, New York 1959.

[Bech-93]   Bechert, D.W.: Verminderung des Strömungswiderstandes durch bionische Oberflächen. In: VDI-Technologieanalyse Bionik, S. 74 – 77. VDI-Technologie-zentrum Düsseldorf 1993.

[Bech-97]   Bechert, D.W., Biological Surfaces and their Technological Application. 28th AIAA Fluid Dynamics Conference: 1997

[DUB-95]   Dubbel, Handbuch des Maschinenbaus, Springer Verlag Berlin, 15.Auflage 1995.

[Eppl-90]   Richard Eppler: Airfoil Design and Data. Springer, Berlin, New York 1990.

[Fel-23] Felgenhauer, Mi. (2023) Fluid-Filament-Wechselwirkung (FFWW). Vortex-SuPerformance. Grin-Verlag München, ISBN: 9783346804464, V1321098

[Fel 22-6]   Felgenhauer, Mi. (2022). Fluid within Fluid Modellierung. Methoden für das FwF Computing. GRIN-Verlag GmbH München, ISBN(e-Book): 9783346662620; ISBN (Buch):9783346662637, VNR: v1234590

[Fel 22-5]   Felgenhauer, Mi. (2022). Modelle und Simulation synthetischer Wirbelspulen. GRIN-Verlag GmbH München, ISBN (eBook): 9783346656353, VNR: v1225465

[Fel 22-1]   Felgenhauer, Mi. (2022). Zur Fluids within Fluid Phänomenologie. Thoughts on a FwF Phenomenology. GRIN-Verlag GmbH München, ISBN(e-Book): 9783346595799 ISBN (Buch): 9783346595805, VNR: v1172544

[Gorr-17]   Edgar Gorrell, S. Martin: Aerofoils and Aerofoil Structural Combinations. In: NACA Technical Report. Nr. 18, 1917.

[Hal-10]   G. Haller. (2010) A variational theory of hyperbolic Lagrangian Coherent Structures. Physica D: Nonlinear Phenomena,240(7):574–598,2010.

[Hal-00]   G. Haller, G.Yuan Lagrangian coherent structures and mixing intwo-dimensional turbulence, Division of Applied Mathematics, Lefschetz Center for Dynamical Systems, Brown University, Providence, RI 02912, USA Received 11 February 2000;

[Hal-01]   Haller, G. (2001). Distinguished material surfaces and coherent structures in three-dimensional fluid flows. Physica D: Nonlinear Phenomena, 149(4),

[Hal - 05] Haller, G. (2005) An objective definition of a vortex J. Fluid Mech. 525, 1-26.

[Hal-11]   Haller, G. (2011). A variational theory of hyperbolic Lagrangian coherent structures. Physica D: Nonlinear Phenomena, 240(7),
Haller, G. (2015). Lagrangian coherent structures. Annual Review of Fluid Mechanics,

[Hal-14]   Farazmand, M., Blazevski, D., & Haller, G. (2014). Shearless transport barriers in unsteady two-dimensional flows and maps. Physica D: Nonlinear Phenomena, ESSOAr

[Hal - 13] Haller, G., & Beron-Vera, F. J. (2013) Coherent Lagrangian vortices: The black holes of turbulence. J. Fluid Mech., 731, R4, 2013.

[Hal – 15-1] Haller, G. (2015) Lagrangian Coherent Structures. Annual Rev. Fluid. Mech, 47, 137-162.

[Hal – 15-2] Haller, G. (2015) Dynamically consistent rotation and stretch tensors for finite continuum deformation. submitted.

[Hal-16]   Haller, G., Hadjighasem, A., Farazmand, M., & Huhn, F. (2016). Defining coherent vortices objectively from the vorticity. Journal of Fluid Mechanics, 795

[Hel - 1858] Helmholtz, H. (1858) über Integrale der hydrodynamischen Gleichungen, welche den Wirbelbewegungen entsprechen. J. Reine und Angew. Math. 55, 25-55.
[Hüt-07]   Hütte, 2007, 33. Auflage, Springer Verlag. S.E147

[Hus - 86] Hussain, A. K. M. F. (1986) Coherent structures and turbulence. J. Fluid Mech. 173, 303

[Hun - 88] Hunt, J. C. R., Wray, A. A. & Moin, P. (1988) Eddies, stream, and convergence zones in turbulent flows. Center for Turbulence Research Report CTR-S88, pp. 193{208

[Kar-35]   Karman von, T. Burgess J.M. (1935) General aerodynamic theory: perfect fluids, In Aerodynamic Theory vol. II (cd. W. F. Durand), p. 308. Leipzig: Springer Verlag.

[Katz-01]   Joseph Katz, Allen Plotkin (2001) Low-Speed Aerodynamics (Cambridge Aerospace Series) Cambridge University Press; 2 edition (February 5, 2001)

[Kab-89] Kaschub, M. (1989) Beitrag zur aerodynamischen Leistungsregelung des Wirbelspulen-Windenergie-Konzentrators. Fortschrittsberichte VDI Reihe 7 Nr. 163, VDI-Verlag Düsseldorf.

[Kra-86] Krasny, R. (1986) Desingularization of Periodic Vortex Sheet Roll-up. Courant Instirute oJ' Mathematical Sciences, New York Unioersity, 251Mercer Street, Nen, York, New York 10012, received November 15, 1981; revised July 25, 1985

[Kat-19]   Katsanoulis, S., Farazmand, M., Serra, M., & Haller, G. (2019). Vortex boundaries as barriers to diffusive vorticity transport in two-dimensional flows. arXiv preprint arXiv:1910.07355 .

[Ker-17]   Kern, M., Hewson, T., Sadlo, F., Westermann, R., & Rautenhaus, M. (2017). Robust detection and visualization of jet-stream core lines in atmospheric flow. IEEE transactions on visualization and computer graphics, 24(1), 893{902.

[Lech-14] Lecheler, S. (2014) Numerische Strömungsberechnung Springer Verlag Berlin Heidelberg. ISBN 978-3-658-05201-0

[Lun-82]T. S. Lundgren, T.S. (1982) Strained spiral vortex model for turbulent fine structure, The Physics of Fluids 25, 2193 (1982); https://doi.org/10.1063/1.863957

[McW - 84] McWilliams, J. C., (1984) The emergence of isolated coherent vortices in turbulent flow. Fluid Mech. 146, 21-43.

[McW - 84] McWilliams, J. C., 1984 The emergence of isolated coherent vortices in turbulent flow. Fluid Mech. 146, 21-43

[Mial-05]   B. Mialon, M. Hepperle: "Flying Wing Aerodynamics Studies at ONERA and DLR", CEAS/KATnet Conference on Key Aerodynamic Technologies, 20.-22. Juni 2005, Bremen.

[Mof-84] Moffatt, K.H. (1984) Simple topological aspects of turbulent vorticity dynamics In: Turbulence and Chaotic Phenomena in Fluids, ed. T. Tatsumi (Elsevier) 223-230.

[Nac-01]   Nachtigall, W. (2001) Biomechanik. Braunschweig: Vieweg Verlag.

[Nach-98]   Nachtigall, W. : Bionik – Grundlagen und Beispiele für Ingenieure und Naturwissenschaftler. Springer-Verlag, Berlin-Heidelberg-New York 1998.

[Nach-00]   Nachtigall, Werner; Blüchel, Kurt. Das große Buch der Bionik. Stuttgart: Deutsche Verlags Anstalt: 2000.

[Oert-11]   Oertel jr., H., Böhle, M., Reviol, Th. (2011) Strömungsmechanik, Grundlagen.Springer Verlag Berlin Heidelberg. ISBN 978-3-8348-8110-6

[Pei-89] Peintinger, G. (1989) Theoretische und experimentelle Untersuchungen an einem Segelflügel-Windkonzentrator. Dissertation FB10 Technische Universität Berlin 1989.

[Rech-94]   Rechenberg, Ingo. Evolutionsstrategie'94. Frommann-Holzoog Verlag. Stuttgart: 1994.

[Scha-13]   Schade, H. (2013) Strömungslehre. De Gruyter Verlag. ISBN-13: 978-3110292213

[Sun-16]   Sun,P.N., Colagrossi, A. Marrone, S. , Zhang, A.M, (2016) Detection of Lagrangian Coherent Structures in the SPH framework, College of Shipbuilding Engineering, Harbin Engineering University, Harbin 150001, China; CNR-INSEAN, Marine Technology Research Institute, Rome, Italy; Ecole Centrale Nantes, LHEEA Lab. (UMR CNRS), Nantes, France.

[Tham-08]   Siekmann, H.E., Thamsen, P. U. (2008) Strömungslehre Grundlagen, Springer Verlag Berlin Heidelberg. ISBN 978-3-540-73727-8

[Tho-59]   Thompson, D'Arcy, W. (1959) On Growth and Form. London: Cambridge University Press. (Neuauflage der Originalschrift 1907)

[Tho-92]   Thompson, D W., (1992). On Growth and Form. Dover reprint of 1942 2nd ed. (1st ed., 1917). ISBN 0-486-67135-6

[Tria-92]   Triantafyllou M., Triantafyllou G. S., Grosenbaugh M. A. (1992) Optimal Thrust Development in Oscillating Foils with Application to Fish Propulsion, Journal of Fluids and Structures (Accepted for Publication)Google Scholar

[Vos-15-2]   M. Voß, H.-D. Kleinschrodt, Mi. Dienst: "Experimentelle und numerische Untersuchung der Fluid-Struktur-Interaktion flexibler Tragflügelprofile", Resarch Day 2015 - Stadt der Zukunft Tagungsband - 21.04.2015, Mensch und Buch Verlag Berlin, S. 180- 184, Hrsg.: M. Gross, S. von Klinski, Beuth Hochschule für Technik Berlin, September 2015, ISBN:978-3-86387-595-4.

[Vos-15-1]   M. Voss, P.U. Thamsen, H.-D. Kleinschrodt, Mi. Dienst (2015): "Experimeltal and numerical investigation on fluid-structure-interaction of auto-adaptive flexible foils", Conference on Modelling Fluid Flow (CMFF'15), Budapest, Ungarn, 1.-4. September 2015, ISBN (Buch): 978-963-313-190-9.

[Vos-15-2]   M. Voss, (2015) Experimentelle und numerische Untersuchung flexibler Tragflügelprofile. Dissertation, Technische Universität Berlin 2015.

[Zie - 72]   Zierep, J. (1972) Ähnlichkeitsgesetze und Modellregeln der Strömungslehre.

Attachment

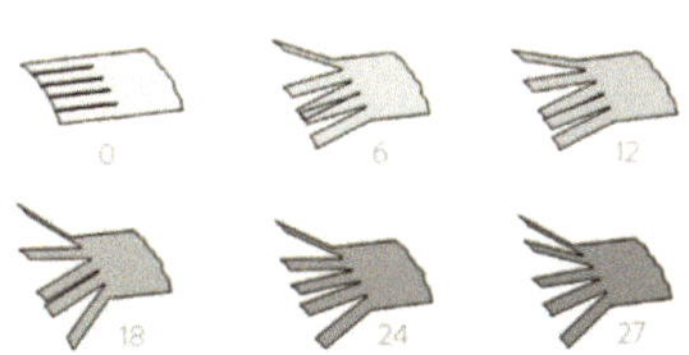

Fig.1: Conditioning of the segmented tip contour of a model wing. Lift L and drag W were measured. After only 27 variation and selection campaigns, the synthetic wing resembles its biological model: the plumage of the land-soaring bird.

Konditionierung der aufgegliederten Randbogenkontur eines Modellflügels. Gemessen wurden Auftrieb L und Widerstand W. Nach nur 27 Variations- und Selektionskampagnen ähnelt der synthetische Flügel seinem biologischen Vorbild: dem Gefieder des landsegelnden Vogels.
Nach Abbildungen aus dem Fundus des Fachgebiets der Bionik und Evolutionstechnik der Technischen Universität Berlin; siehe auch: Stache, M.: Evolutionsstrategisches Design von Tragflügelspitzen. DGLR- Jahrbuch, DGLR-2006-200, 2006, pp. 1131-1138

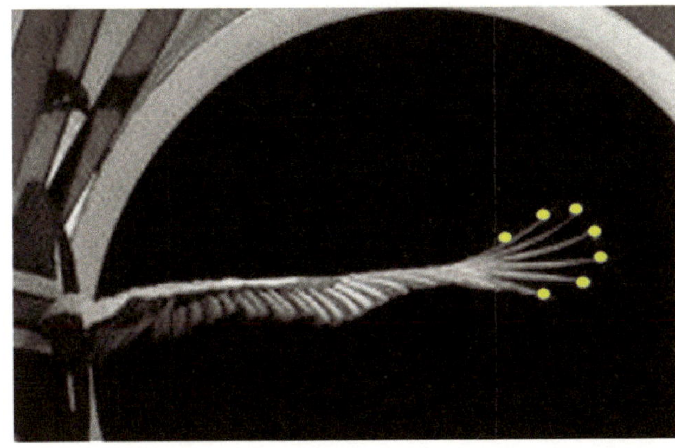

Fig.2: Specimen of the stork wing in the wind tunnel under flow load. The seven source points at the feather finger tips are marked in yellow.
Photography: Fachgebiet Bionik und Evolutionstechnik, Technische Universität Berlin (um 1995).

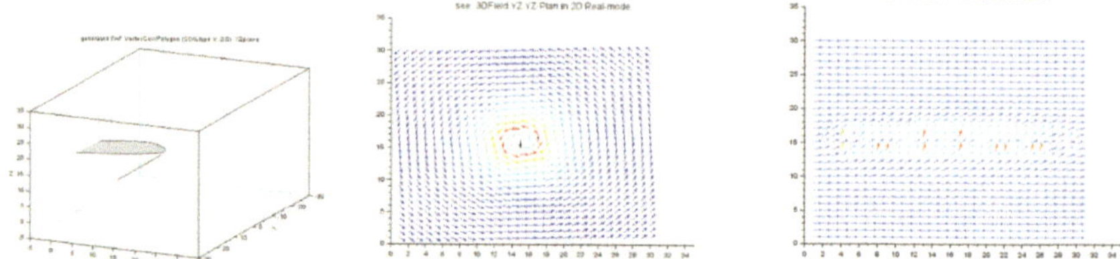

Fig.3: Numerical simulation of a vortex track in a fluid at rest. Schematic: wing and corridor model (left); Distribution of the velocity vectors in the YZ plane (middle) and in the XY plane, on the right in the picture. (nach: Felgenhauer (2023). Fluid-Filament- Wechselwirkung, FFW).

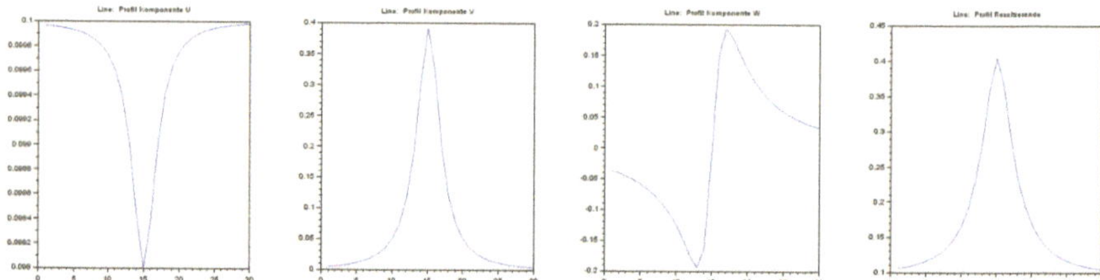

Fig.4: Y-Z section (X=15LE) and the components of the velocity gradients in the direction of the axis of the vortex thread: u; radial components, v, w; Resulting local velocity in the field: r. (nach: Felgenhauer (2023). Fluid-Filament- Wechselwirkung, FFW).

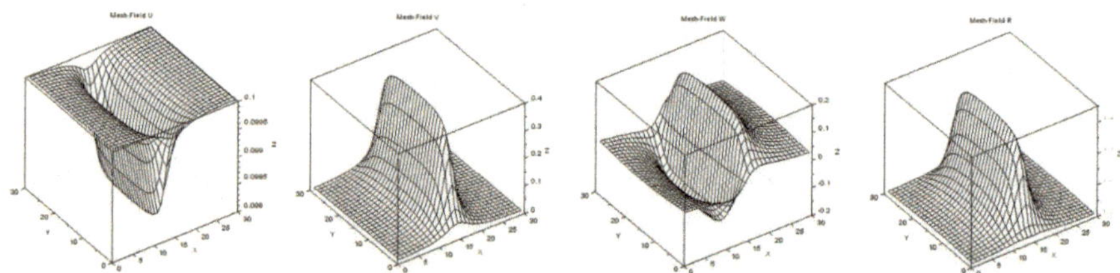

Fig.5: Section through the calculated field in the X-Y plane (Z=15LE). Impulse effect of an "elongated" vortex filament in a fluid at rest (Fig. 3); the components of the velocity gradient u, v, w; and the resultant r. (left to right; nach: Felgenhauer (2023). Fluid-Filament- Wechselwirkung, FFW).

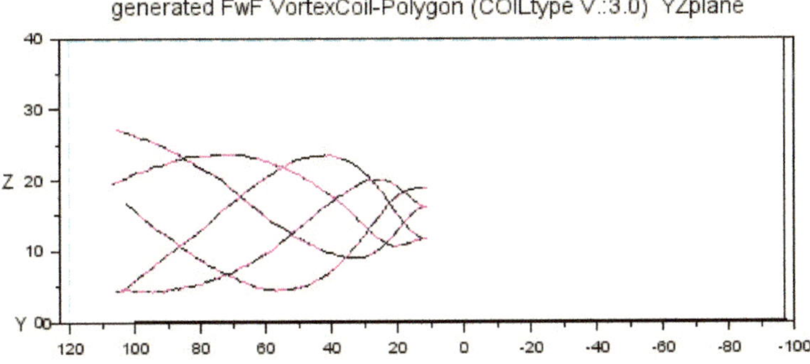

Abb.6: Corridor model of a 5-fold vortex coil: globaMode5WSP.

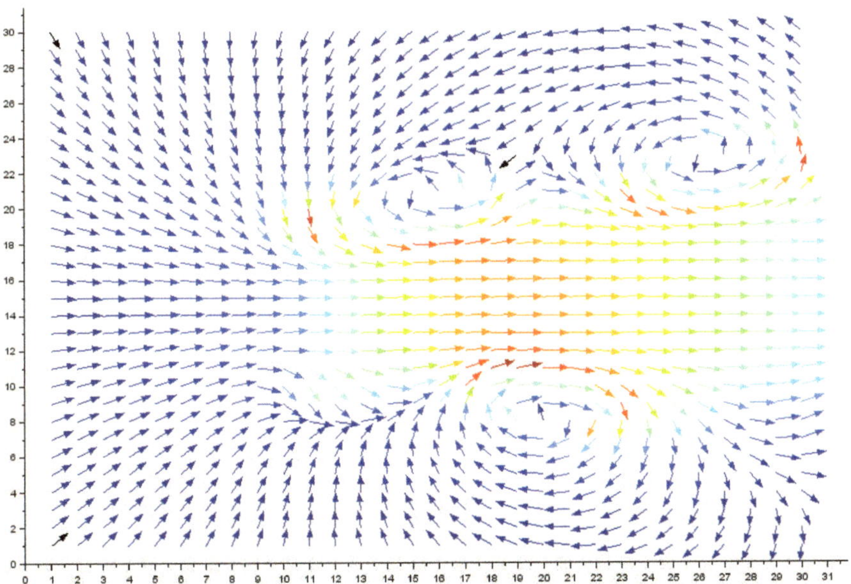

Fig.7: Flow pattern in the XY section plane (Z=15 LE)

YOUR KNOWLEDGE HAS VALUE

- We will publish your bachelor's and
  master's thesis, essays and papers

- Your own eBook and book -
  sold worldwide in all relevant shops

- Earn money with each sale

Upload your text at www.GRIN.com
and publish for free